Q-CODES

A QUICK REFERENCE GUIDE

By

JOHN PERTELL

KC2TAV

"...In the new era, thought itself will be transmitted by radio." ~ Guglielmo Marconi

QRZ?

TABLE OF CONTENTS

HAM RADIO JARGON

Legal Notes

Although the author and publisher have made every effort to ensure that the information in this book was correct at press time, the author and publisher do not assume and hereby disclaim any liability to any party for any loss, damage, or disruption caused by errors or omissions, whether such errors or omissions result from negligence, accident, or any other cause.

This Book is for informational purposes only.

THE HAM RADIO Q-CODES

Commonly Used Q-Codes

QRL

Question: Are you busy?
Response: I am busy. (or I am busy with ...) Please do not interfere.

QRM

Question: Are you being interfered with?
Response: I am being interfered with.

QRN

Question: Are you troubled by static?
Response: I am troubled by static.

QRO

Question: Should I increase power?
Response: Increase power

QRP

Question: Shall I decrease power?
Response: Decrease power

QRQ

Question: Shall I send faster?
Response: Send faster (... wpm)

QRS

Question: Should I send more slowly?
Response: Send more slowly (... wpm)

QRT

Question: Should I stop sending?
Response: Stop Sending

QRU

Question: Have you anything for me?
Response: I have nothing for you

QRV

Question: Are you ready?
Response: I am ready.

QRX

Question: When will you call me again?
Response: I will call you again at ... (hours) on ... khz (or Mhz)

QRZ

Question: Who is calling me?
Response: You are being called by ... on ... khz (or Mhz)

QSB

Question: Are my signals fading?
Response: Your signals are fading.

QSL

Question: Can you acknowledge receipt?
Response: I am acknowledging receipt

QSO

Question: Can you communicate with ... direct or by relay?
Common Useage -- A 2-Way Contact
Response: I can communicate with ... direct (or by relay through ...).

QSX

Question: Will you listen to ... (call sign(s) on ... khz (or Mhz))?
Response: I am listening to ... (call sign(s) on ... khz (or Mhz))
Commonly used on the DX Packet Clusters to indicate where the DX station was listening or contacted during a split operation

QSY

Question: Shall I change to transmission on another frequency?
Response: Change to transmission on another frequency (or on ... khz (or Mhz)).

QTH

Question: What is your position in latitude and longitude (or according to any other indication)?
Where is your station located?
Response: My position is ... latitude...longitude, I am located at…

COMPLETE LIST OF THE HAM RADIO Q-CODES *(INCLUDING COMMONLY USED)*

(= COMMONLY USED)*

QCB

Question: Delay
Response: Delay is being caused by ...

QCS

Question: Frequency Breakdown
Response: My reception on ... frequency has broken down.

QCX

Question: What is your full call sign?
Response: My full call sign is ...

QDP

Question: Have you sent message... to ...?
Response: I have sent message ... to ...

QIF

Question: What frequency is ... using?
Response: ... is using ... khz (or ... mhz.).

QMH

Question: Shift Frequency

Response: Shift to transmit and receive on ... khz. (or ... mhz.); if communication is not established within 5 minutes, revert to present frequency.

QRA

Question: What is the name of your station?
Response: The name of my station is ...

QRB

Question: How far are you from my station?
Response: I am ____ km from you station

QRG

Question: Will you tell me my exact frequency (or that of ...)?
Response: Your exact frequency (or that of ...) is ... khz (or Mhz).

QRH

Question: Does my frequency vary?
Response: Your frequency varies.

QRI

Question: How is the tone of my transmission?
Response: The tone of your transmission is (1. Good; 2. Variable; 3. Bad).

QRJ

Question: Are you receiving me badly?
Response: Are you receiving me badly?

QRK

Question: What is the readability of my signals (or those of ...)?
Response: The readability of your signals (or those of ...) is ... (1 to 5).

* QRL

Question: Are you busy?
Response: I am busy. (or I am busy with ...) Please do not interfere.

*QRM

Question: Are you being interfered with?
Response: I am being interfered with.

*QRN

Question: Are you troubled by static?
Response: I am troubled by static.

*QRO

Question: Should I increase power?
Response: Increase power

*QRP

Question: Shall I decrease power?
Response: Decrease power

*QRQ

Question: Shall I send faster?
Response: Send faster (... wpm)

*QRS

Question: Should I send more slowly?
Response: Send more slowly (... wpm)

*QRT

Question: Should I stop sending?
Response: Stop Sending

*QRU

Question: Have you anything for me?
Response: I have nothing for you

*QRV

Question: Are you ready?
Response: I am ready.

QRW

Question: Should I inform ... that you are calling him on ... khz (or Mhz)?
Response: Please inform ... that I am calling him on ... khz (or Mhz)

*QRX

Question: When will you call me again?
Response: I will call you again at ... (hours) on ... khz (or Mhz)

*QRZ

Question: Who is calling me?
Response: You are being called by ... on ... khz (or Mhz)

QSA

Question: What is the strength of my signals (or those of ...)?
Response: The strength of your signals (or those of ...) is ... (1 to 5).

**QSB*

Question: Are my signals fading?
Response: Your signals are fading.

QSD

Question: Is my keying defective?
Response: Your keying is defective

QSI

Question: Unable To Break in
Response: I have been unable to break in on your transmission. Or .. Will you inform ... (callsign) that I have been unable to break in on his/her transmission on ... khz (or Mhz.)

QSK

Question: Can you hear me between your signals?
Response: I can hear you between my signals.

**QSL*

Question: Can you acknowledge receipt?
Response: I am acknowledging receipt

QSM

Question: Should I repeat the last telegram (message) which I sent you, or some previous telegram (message)?
Response: Repeat the last telegram (message) which you sent me (or telegram(s) (message(s)) numbers(s) ...).

QSN

Question: Did you hear me (or ... (call sign)) on .. khz (or Mhz)?
Response: I did hear you (or ... (call sign)) on ... khz (or Mhz).

**QSO*

Question: Can you communicate with ... direct or by relay?
Common Useage -- A 2-Way Contact
Response: I can communicate with ... direct (or by relay through ...).

QSU

Question: Shall I send or reply on this frequency (or on ... khz (Mhz)) (with emissions of class ...)?
Response: Send or reply on this frequency or(or on ... khz (Mhz)) (with emissions of class ...)

QSV

Question: Shall I send a series of V's on this frequency (or ... khz (or Mhz))?
Response: Send a series of V's on this frequency (or ... khz (or Mhz))

QSW

Question: Will you send on this frequency (or on ... khz (or Mhz)) (with emissions of class ...)?
Response: I am going to send on this frequency (or on ... khz (or Mhz)) (with emissions of class ...)

*QSX

Question: Will you listen to ... (call sign(s) on ... khz (or Mhz))?
Response: I am listening to ... (call sign(s) on ... khz (or Mhz))
Commonly used on the DX Packet Clusters to indicate where the DX station was listening or contacted during a split operation

*QSY

Question: Shall I change to transmission on another frequency?
Response: Change to transmission on another frequency (or on ... khz (or Mhz)).

QSZ

Question: Shall I send each word or group more than once?
Response: Send each word or group twice (or ... times)

QTA

Question: Shall I cancel telegram (message) No. ... as if it had not been sent?
Response: Cancel telegram (message) No. ... as if it had not been sent.

QTG

Question: Will you send two dashes of ten seconds each followed by your call sign (repeated ... times) (on ... khz (or Mhz))? Or Will you request ... to send two dashes of ten seconds followed by his call sign (repeated ... times) (on ... khz (or Mhz))?
Response: I am going to (or I have requested ... to) send two dashes of ten seconds followed by call sign (repeated ... times) (on ... khz (or Mhz))

**QTH*

Question: What is your position in latitude and longitude (or according to any other indication)?
Where is your station located?
Response: My position is ... latitude...longitude, I am located at…

QTQ

Question: Can you communicate with my station by means of the International Codes of Signals?
Response: I am going to communicate with your station by means of the International Codes of Signals

QTR

Question: What is the correct time?
Response: The correct time is ... hours

QTS

Question: Will you send your call sign for ... minutes(s) now (or at ... hours) (on ... khz (or Mhz)) so that your frequency may be measured?

Response: I will send my call sign for ... minutes(s) now (or at ... hours) (on ... khz (or Mhz)) so that my frequency may be measured

QTU

Question: What are the hours during which your station is open?
Response: My station is open from ... to ... hours.

QTV

Question: Should I monitor for you on the frequency of ...khz (or Mhz) (from ... to ... hours)?
Response: Monitor for me on the frequency of ...khz (or Mhz) (from ... to ... hours)

QTX

Question: Will you keep your station open for further communication with me until further notice (or until ... hours)?
Response: I will keep my station open for further communication with you until further notice (or until ...hours)

QUC

Question: What is the number (or other indication) of the last message you received from me (or from .. (call sign))?
Response: The number (or other indication) of the last message I received from you (or from ... (call sign) is ...

HAM RADIO JARGON

A - "ALPHA"

Absorption –
The reduction of radio signal strength in the Ionosphere.

AC –
Alternating Current

Access Code –
A code to activate a repeater function such as a Link or an Auto Patch.

A/D –
Analog-to-Digital

Aerial –
An outdoor antenna.

AF –

Audio Frequency. Humans hear from 20 to 20,000 Hertz.

AFC –
Automatic Frequency Control.

AFSK –
Audio Frequency Shift Keying

A-index –
The index of the Earth's magnetic field.

AGC –
Automatic Gain Control. Prevents fading in a receiver.

ALC –
Automatic level control. Prevents amplifier overload in a transmitter's output.

Amateur-
A person who is licensed to operate in the amateur radio bands.

Amateur Radio –
Non-commercial radio. Rules for operating Amateur Radio in the United

States are listed under Part 97 of the FCC Rules and Regulations.

Ampere – (Amps)
A unit of electrical "Current".

AM –
Amplitude Modulation

AMSAT –
Radio Amateur Satellite Corporation

AMTOR –
Amateur Teleprinter Over Radio.

Antenna -
A device that radiates or captures radio frequency.

Antenna Farm –
An area containing many Ham Radio antennas.

Antenna Tuner –
A device that matches the antenna system's input impedance to the transceiver's output impedance.

APRS –
Automatic Packet Position Reporting System.

ARC –

Amateur Radio Club.

ARES –
Amateur Radio Emergency Service.

ARRL –
American Radio Relay League

ARQ –
Automatic repeat request. Used in AMTOR.

ASCII –
American Standard Code for Information Interchange.

ASR –
Automatic Send-Receive. Used in RTTY to allow message composition while receiving text from another station.

ATT –
Attenuator. Expressed in reduction of Decibels, or "DB".

Auto patch –
An interface from a repeater to a telephone system that allows users to make phone calls through the repeater.

AVC –

Automatic Volume Control. It levels out a receiver's audio volume.

AWG –

American Wire Gauge. A number system that describes the diameter of a wire. As the numbers decrease, the diameter of the wire increases.

B –"BRAVO"

Balanced Line –

A feed line with two conductors having equal but opposite voltages. Both with neither conductor at ground potential.

Balanced Modulator –

A mixer circuit used in a single-sideband suppressed-carrier transmitter to combine a voice signal and the RF carrier. The balanced modulator isolates the input signals from each other and the output, so that only the sum and the difference of the two input signals reach the output. The original carrier signal and the audio signal are suppressed.

Balun –

Balanced to unbalanced. The device couples a balanced antenna to an unbalanced feed line.

Band –

A range of frequencies allocated for a particular use.

Band pass –

A range of frequencies permitted to pass through a receiver circuit or filter.

Band-pass filter –

A circuit that allows a specific range of frequencies to pass through - yet attenuates signals below and above this range.

Base –

A radio station that is at a fixed location as opposed to a mobile.

Barefoot –

Transmitting with a transceiver alone. A linear amplifier is not used.

Base Loading –

The loading coil is placed at the bottom of the antenna This accomplishes a lower resonant frequency.

BAUD –

The unit of digital-signal speed.

BCI –

Broadcast Radio Interference.

Beam –
An antenna that radiates directionally.

Beacon –
A station that transmits one-way signals for the sole purpose of homing, navigation and determining propagation conditions.

BFO –
Beat frequency oscillator. It is used to mix with the incoming signal to produce an audio tone for CW reception.

Bird –
It is a nickname for a satellite.

Birdie –
These are spurious signals sometimes produced in a receiver – They usually are the product of mixed intermediate frequencies occurring within the radio.

Bleed Over-
Interference that is caused by a station operating on an adjacent frequency.

Bleeder Resistor –
A large Ohm value resistor that connects across the filter capacitor in a power supply. Used to discharge the filter capacitors when the supply is turned off.

Block Diagram –

A drawing that uses rectangles to represent sections of electronic circuits. The block diagram shows signal flow and the function of the sections.

BNC –

Bayonet Niell-Concelman – Named after it's inventors. It is a coax connector commonly used with VHF/UHF equipment.

BPSK –

Binary Phase Shift Keying. Digital DSB suppressed carrier modulation.

Boat Anchor –

A reference to antique Ham Radio equipment. Named because antique radios were large and heavy.

Bounce –

The "bouncing" or reflection of a radio wave off of an object during transmission.

Breadboard –

An experimental board to test electronic circuits. Named as such because early experimenters used a wood board or bread board to lay out circuits.

Broadcasting –

Transmissions intended for the general public. Broadcasting is prohibited on the Amateur Radio Bands.

Bug –

It is a semi-automatic mechanical code key

C – "CHARLIE"

California Kilowatt –
Transmitting power set to a level above the legal limit.

Calling Frequency –
A set frequency where stations attempt to contact each other.

Candy Store –
A term for a Ham Radio Dealer.

Cans –
A set of headphones.

CAP –
The Civil Air Patrol

Capacitor –
Sometimes it is referred to as a "Cap". It is an electronic component that is made up of two or more

conductive plates separated by a nonconductive material. A capacitor stores electricity.

Carrier –
It is a fixed radio frequency emission without modulation or interruption. Modulation can later be applied to the carrier, such as AM (Amplitude Modulation) or FM (Frequency Modulation).

CATV –
Cable Television

CATVI –
Cable Television Interface.

CBA –
Call book Address

CCW –
Coherent CW

Center Frequency –
An FM transmitter's unmodulated carrier frequency.

Center Loading –
A loading coil is placed at the center of an antenna. This is used to achieve a lower resonant frequency.

Channel –

It is the pair of input and output frequencies used by a repeater.

Chassis Ground –

It is the common ground of all the parts of the circuit that join together at the negative side of the power supply.

Chirp –

A chirping sound caused by changes in the carrier frequency of a CW transmitter.

Clear –

It is a statement used to indicate that a station is done transmitting.

Closed Repeater –

The access to the Repeater is limited to a select group.

Cloud Warmer –

This is an expression used for an antenna which radiates nearly straight up.

CMOS –

Complementary Symmetry Metal Oxide emiconductor.

Coax / Coaxial Cable –

A type of wire that contains a center wire surrounded by insulation and then a grounded shield of braided wire. The shield minimizes radio frequency and electrical interference. While there are different impedances to coax, 50-ohm is the most commonly used in Ham Radio.

Code –
A term that usually refers to Morse code. However, it can also be used for others such as baudot.

Coil –
A conductor that is wound into a series of loops.

Color Code –
Numerical values are represented by specific colors and then used to identify the value of certain electrical components. Such as resistors.

Condenser –
This is an old term for a capacitor.

Controller –
It is the system that controls the functions of the repeater. Such as turning the repeater on and off, controlling auto patch, setting CTCSS tones ect.

Control Operator –
As required by FCC regulations, this is the Amateur Radio operator designated to "control" the operation of the repeater.

Copy –
This is a term to indicate how well communication is received.

Copying –
Used to indicate that one is monitoring in on a conversation.

Core –
The material used in the center of an inductor coil, where the magnetic field is concentrated.

Courtesy Beep –
An audible sound that is produced to indicate a user may go ahead and transmit on the repeater.

Coverage –
The geographic area that is covered by a repeater.

CPS –
Cycles Per Second. Later, this terminology was replaced by "Hertz".

CQ –
Calling any amateur radio station. Can be used in all modes of transmission.

Critical Angle –

It is the angle at which a radio signal is refracted in the ionosphere. Generally, the higher the angle, the shorter the distance. The lower the angle the greater the distance of transmission.

Critical Frequency -
The highest frequency at which a vertically incident radio wave will return from the ionosphere. Above the critical frequency radio signals pass through the ionosphere instead of returning to Earth.

Cross Band –
The process of transmitting on one band and having it receive another.

CRT –
Cathode Ray Tube

Crystal –
A piezoelectric device that resonates at a frequency dependent on its material, dimensions, and temperature.

Crystal Filter –
A network of piezoelectric crystals used for the rejection of unwanted signals.

Crystal Oscillator –
It is a device or component that uses a quartz crystal to keep the frequency of a transmitter constant.

CSCE –

In the U.S., this is the "Certificate of Successful Completion of Examination". It is the certificate certifying a person has successfully passed one or more of the amateur radio license examinations.

CTCSS –

Continuous Tone Controlled Squelch System. Sub audible tones that may be used by repeaters to block access. The repeater will not transmit unless the proper CTCSS tone is received.

Current –

It is the flow of electrons in an electrical circuit.

Cutoff Frequency –

The frequency at which a filter will begin to reject signals.

CW –

Continuous Wave. Morse Code Transmissions.

D – "DELTA"

dB –
A Decibel is $1/10^{th}$ of a Bell. A unit used to measure the intensity of a sound or the power level of an electrical signal.

dBc –
Decibels relative to the carrier level.

dBd –
Decibels above or below a dipole antenna.

dBi –
Decibels above or below an isotropic antenna.

DC –
Direct current.

DE –
Morse code for "from".

Delta Loop Antenna –
Same principal as a Cubical Quad antenna except with triangular elements.

Desensitization -
Overload is caused from a nearby transmitter. This causes the receiver to lose receiving sensitivity.

Detector –
It is the stage in a receiver in which the modulation is recovered from the RF signal. In FM this is called the discriminator.

Deviation –
It is the change of an FM transmitter's carrier frequency caused by the modulating signal.

Deviation Ratio –
Used in an FM transmitter. It is the ratio between the maximum change in RF carrier frequency and the highest modulating frequency.

Digipeater –
A digital repeater which receives and transmits data packets on the same frequency.

Dip Meter –

A device that determines the resonant frequency of an electronic circuit.

Diplexer -

Typically used to couple two transceivers to a single or dual band antenna. This allows one to receive on one transceiver and transmit on the other transceiver.

Dipole –

An antenna that is open and fed at the center. The entire antenna is ½ wavelength long at the desired operating frequency.

Director –

It's the element located in front of the driven element in directional antennas.

Doubling –

When two stations transmit simultaneously, the signals mix in the repeater's receiver. This results in a scratchy sounding signal. FM has a characteristic whereby the stronger signals captures and over rides the weaker one.

Downlink –

Channel used for satellite-to-earth communications.

DPDT -

Double Pole, Double Throw Switch. Switches two different circuit lines to Two different points.

DPST –
Double Pole, Single Throw Switch. Switches two different circuit lines on or off.

DPSK –
Differential Phase Shift Keying. This is a form of BPSK where only data transitions are transmitted.

D-layer –
This is the lowest region of the ionosphere and found approximately 25 to 55 miles above Earth. It quickly fades away after sunset and sometimes does not form at all on short winter days. The D-Layer impacts radio propagation by absorbing energy from the signals that pass through it.

Driven Element –
This is the antenna element that connects directly to the feed line.

Dropping Out -
Repeaters require a minimum signal in order to transmit. When a signal does not have enough power to keep the repeater transmitting, it stops transmitting, or "drops out".

DSP –

Digital Signal Processing. Allows for filtering, noise reduction, audio equalization.

DTMF –

Dual Tone Multi Frequency. It's the series of tones generated from a keypad on a ham radio transceiver.

Dual Band Antenna -

This is an antenna that can be used on two different radio bands. An example would be 2 meter/ 70 cm.

Dummy Load –

This is a device which substitutes for an antenna during tests on a transmitter. It converts radio energy to heat instead of radiating energy.

Duplex -

This is a communication mode in which a radio transmits on one frequency and receives on another.

Duplexer –

This is a device used in repeater systems that allow a single antenna to transmit and receive simultaneously.

DVM –

Digital voltmeter

DX –

As a Noun - A distant station; If being used as a Verb - to contact a distant station.

DXer –

An Amateur radio operator who pursues contacting distant and rare Amateur Radio stations.

DXpedition –

This is a term for a radio expedition to remote and rare locations.

Dynamic Range -

How well a receiver can handle strong signals without overloading. Over 100 decibels is considered excellent.

– "ECHO"

Earth Ground –

A circuit connection that terminates to a ground rod. The rod is usually driven into the earth.

Echolink –

Uses a network protocol called VoIP (Voice over IP). This program allows worldwide connections to be made between stations, from computer to station, or from computer to computer. For Repeaters, and Echolink repeater is needed.

ECSSB –

Exalted-carrier single sideband.

EEPROM –

Electrically Erasable Programmable Read Only Memory

E-layer –

The region of the ionosphere found approximately 55 to 90 miles above Earth. This region fades away a few hours after sunset. The main impact of the E-layer on radio propagation is to absorb energy from signals passing through it, although sporadic-E propagation makes possible distant communications on frequencies above 30 MHz.

EHF –

Extremely High Frequency (30 - 300 GHz).

EIRP –

Effective Isotropic Radiated Power.

ELF –

Extremely Low Frequency (30 - 300 Hz)

Elmer –

A mentor for Ham Radio. An experienced operator who tutors newer operators.

Eleven Meters –

This is currently the CB band, or Citizens Band. It was once a Ham band.

EME –

Earth Moon Earth or Moonbounce. Using the moon as a passive reflector to establish a signal path.

EMF –

Electromotive force. “Voltage”.

EMI –
Electromagnetic Interference.

Emission Mode –
The type of radio emission. Such as AM, FM, or Single Sideband.

EMP –
Electromagnetic Pulse. This is an extremely high energy magnetic emission. Such as a lightning strike or nuclear explosion.

ERP –
Effective Radiated Power.

E-skip –
Sporadic E-layer ionospheric propagation.

F – “FOXTROT”

FAA –
Federal Aviation Administration (USA).

F-layer –
The region of the ionosphere found approximately 90 to 400 miles above Earth and which is responsible for most long distance propagation on frequencies below 30 MHz. During the daytime and especially in summer, solar heating can cause the F-layer to split into two separate layers. The F1-layer and the F2-layer.

FAQ –
Frequently Asked Questions

Far Field of an antenna –
That region of the electromagnetic field surrounding an antenna where the field strength as a function of angle (the antenna pattern) is essentially independent of the distance from the antenna. The field in this region also demonstrates more of a plane wave character.

Farnsworth –

This is a method of sending Morse code characters. Example characters are sent at 13 words per minute but the spacing is adjusted so that the overall code speed is 5 words per minute.

Fax –

Facsimile. A digital mode for transmitting images.

FCC –

Federal Communications Commission, the overnmental body in the U.S. which regulates the radio spectrum.

Feed line –

A wire or cable connecting a radio to an antenna.

FET –

Field-effect transistor.

Field Day –

This is an Amateur Radio activity that usually occurs in June to practice emergency communications.

Field Strength Meter –

This is a meter used to show the presence of RF energy and the relative strength of the RF field.

Filter –

A circuit or device that will allow certain frequencies to pass while rejecting others.

Final –

This can be either the Last amplifying stage of a radio transmitter, or the last transmission by a station during a contact.

Flat Topping –

Over modulating so as to distort a waveform.

Flutter –

Rapid variation in the signal strength of a station, usually due to propagation variations.

FM –

Frequency Modulation.

Fox Hunt –

This is a contest to locate a hidden transmitter.

Frequency –

The rate of oscillation. Audio and radio wave frequencies are measured in Hertz.

Frequency Coordinator –

This is an individual or group responsible for assigning frequencies to new repeaters without causing interference to existing repeaters.

FSK –

Frequency Shift Keying. This is a frequency modulation scheme in which digital information is transmitted through discrete frequency changes of the carrier wave. In Binary Frequency Shift Keying - a simple form of FSK. 1 is called the "Mark" and 0 is called the "Space".

FSTV –

Fast-Scan TV.

Full-Break In (QSK) -

Allows a station to break into the communication without waiting for the transmitting station to finish.

Full Duplex –

This is a communications mode in which a radios can transmit and receive at the same time by using two different frequencies.

Full Quieting –

A phenomenon on FM transmissions where the incoming signal is sufficient to engage the receiver limiters - thus eliminating the noise due to amplitude fluctuations.

Full Wave Bridge Rectifier –

A full-wave rectifier circuit that uses four diodes and does not require a center-tapped transformer. This converts AC to DC.

Full Wave Rectifier –

This is a circuit basically composed of two half-wave rectifiers. The full wave rectifier allows the full ac waveform to pass through; one half of the cycle is reversed in polarity. This circuit requires a center-tapped transformer. Converts AC to DC.

Fuse –

Usually made of a thin metal strip mounted in a holder. When excessive current passes through the fuse, the metal strip will melt. This then opens and protects the circuit. Fuses are rated in amperes and voltage and time to activate. “Fast blowing or slow blowing”.

G - "GOLF"

Gallon –

This is a term used for transmitter output power. Legally this would be either 1000 watts CW or 1500 watts PEP. A "Full Gallon" would refer to using the maximum legal power limit.

Gain -

Antenna term. This is an increase in the effective power radiated by an antenna in a certain desired direction, or an increase in received signal strength from a certain direction. This is at the expense of power radiated in, or signal strength received from, other directions.

GHz –

Gigahertz. This is a billion (1,000,000,000) Hertz.

GMRS –

General Mobile Radio Service.

GOTA –

Get On The Air. A category in the annual ARRL Field Day event. The GOTA station may be operated by Novice, Technicians or generally inactive hams under their existing operating privileges, or under the direction of a Control Operator with appropriate privileges, as necessary. Non-licensed persons may participate under the direct supervision of an appropriate control operator.

GPS –
Global Positioning System.

Gray Line –
This is a band around the Earth that separates daylight from darkness. It is a transition region between day and night. Another type of propagation path.

Great Circle Route –
The shortest path by radio between any two points on Earth.

Green Stamp –
A U.S. dollar bill is sent along with a QSL card to offset postage costs of a return card.

Grid Dip Meter –
Test device that causes a meter to decrease, or "dip" when near resonant circuits.

Ground –

Common zero voltage reference point.

Ground Plane Antenna –

This is a vertical antenna built with the central radiating element one-quarter-wavelength long and several radials extending horizontally from the base. The radials are slightly longer than one-quarter wave, and may droop toward the ground.

Ground Wave Propagation –

Radio waves that travel along the surface of the earth.

 - "HOTEL"

Half Duplex –
A communications mode in which a radio transmits and receives on two different frequencies but performs only one of these operations at a time.

Half Wave Dipole –
This is a basic antenna consisting of a length of tubing or wire that is open and fed at the center. The entire antenna is ½ wavelength long at the desired operating frequency.

Half Wave Rectifier –
This is a circuit that allows only half of the applied ac waveform to pass through it.

Hand Held –
This is a small, lightweight portable transceiver small enough to be carried. It can also be called an HT or "Handy Talkie".

Hang Time –

this is the short period following a transmission that allows others who want to access the repeater a chance to do so. A courtesy beep may sound when the repeater is ready to accept another transmission.

Ham –

An amateur radio operator.

Hamfest –

Ham Festival. This is a social and commercial event for hams to meet. Used for buying, selling and swapping equipment.

Harmonic –

This is a signal at a multiple of the fundamental frequency.

HDTV –

High Definition Television

HDX –

Half-duplex. A communication system in which stations take turns transmitting and receiving.

Hertz –

This is the standard unit used to measure frequency. One Hertz equals one complete cycle per second.

HF –

High Frequency. 3 MHz to 30 MHz .

Hi Hi –

Ha ha (laughter). Probably taken from the laughing sound produced while using it in its CW form. Later transferring over to voice.

High Pass Filter –

A filter designed to pass high frequency signals, while blocking lower frequency signals.

Homebrew –

This is a term for home-built. Noncommercial radio equipment or software.

Hop –

Communication between stations by reflecting the radio waves off of the ionosphere.

Horizontally Polarized Wave –

This is an electromagnetic wave with its electric lines of force parallel to the ground.

HT –

Handie-Talkie. This is a small hand held radio.

Hz –

An Abbreviation for “Hertz”.

I - "INDIA"

I –

Intensity. Symbol for current in an electric circuit. Measured in Amperes.

IC –

Integrated circuit.

IF –

Intermediate Frequency. Resultant frequency from heterodyning the carrier frequency with an oscillator. Mixing incoming signals to an intermediate frequency enhances amplification, filtering and the processing signals.

Image –

A false signal produced in a superheterodyne receiver's circuitry.

Impedance –

The opposition to the flow of electric current and radio energy. This is measured in ohms, with the symbol represented as a “Z”. For best performance, the impedance of an antenna, the feedline, and the antenna connector on a radio should be approximately equal.

Inductance –

The measure of the ability of a coil to store energy in a magnetic field.

Inductor –

This is an electrical component usually composed of a coil of wire wound on a central core. An inductor stores energy in a magnetic field.

Input Frequency –

The frequency of the repeater's receiver and your transceiver's transmitter frequency.

Intermod –

Short for "intermodulation". False or spurious signals that are produced by two or more signals mixing in a receiver or repeater station.

Intermodulation Distortion –

Also known as “IMD”. This is the unwanted mixing of two strong RIF signals that cause a signal to be transmitted on an unintended frequency.

I/O –

Input/Output.

Ionosphere –
The electrically charged region of the Earth's atmosphere located approximately 40 to 400 miles above the Earth's surface.

IRLP –
Internet Radio Linking Project. It uses a network protocol called VoIP (Voice over Internet Protocol). There are many repeaters around the world that are connected by IRLP using VOIP.

Isotropic –
A theoretical antenna reference point used when calculating gain.

ITU –
International Telecommunications Union. The organization that specifies worldwide guidelines concerning the use of the electromagnetic spectrum for communications purposes.

J Antenna –

Also known as a “J pole.” This style antenna consists of a half-wavelength radiator fed by a quarter-wave matching stub. This antenna does not require the ground plane that ¼-wave antennas do to work properly.

Jam –

Causing interference intentionally.

JFET –

Junction Field Effect Transistor.

Jug –

Large transmitting tubes.

K – "KILO"

Kc –
Kilocycles

K Index –
A measure of the Earth's magnetic field. Propagation conditions improve with lower measurement numbers.

Kerchunking –
Tripping or activating a repeater without identifying or modulating the carrier. Makes a "kerchunk" sound.

Key –
As a noun, it is the switch or button used in Morse code. As a verb, it means to press a key or button.

Keyer –
This is an electronic device for sending Morse Code semi-automatically. It connects to a key and " Dits" are sent by pressing one paddle of the key. The "dahs" are sent by pressing the other paddle.

Key Up –

Turning on a repeater by transmitting the repeater's input frequency.

Kilocycles –

This is a thousand cycles per second. The term was later replaced by kilohertz "kHz".

Kilohertz –

One thousand hertz.

- "LIMA"

Ladder Line –

A wire resembling a "ladder". It has an open wire transmission line. Typically, 600 and 450 ohm impedances are used.

Landline –

Telephone.

LCD –

Liquid Crystal Display.

LED –

Light-Emitting Diode.

LEO –

These are Low Earth Orbiting Satellites. Satellites used in communication systems which orbit about 400 to 1000 miles above Earth's surface. They are fast moving and are not fixed in space in relation to the Earth.

LF –

Low Frequency - 30 kHz to 300 kHz

Lid –
A person who does not follow the proper procedures or is sloppy at CW.

Limiter –
This is the stage of an FM receiver that clips the tops of the FM signal and making the receiver less sensitive to amplitude variations and pulse noise.

Linear –
This is an amplifier used after the transceiver output. Linear, in the mathematical sense, means that what comes out is directly proportional to what goes in. In a Linear Amplifier, if you double the input, the output is doubled and so forth. This does not generate any additional frequency by-products.

Line Of Sight Propagation –
This is a term used to describe propagation in a straight line. From one station directly to another.

Load –
An electrical device which consumes, converts, or emanates energy.

Local Oscillator -

This is a receiver circuit that generates a stable, pure signal used to mix with the received RF to produce a signal at the receiver intermediate frequency (IF).

Long Path –

The longest path a radio signal could take to reach its destination. Sometimes literally around the world.

Lollipop –

This is an Amateur radio term for the Astatic D-104 microphone.

Low Pass Filter –

This is a filter which allows signals below the cutoff frequency to pass through while attenuating the signals above the cutoff frequency.

LSB –

Lower Side Band. The commonly used single sideband operating mode on the 40, 80, and 160 meter amateur bands.

LW –

Long Wave 150 - 300 KHz

- "MIKE"

mA –
Milliampere 1/1,000 ampere.

Machine –
A repeater.

Magnetic Mount –
"Mag Mount". This is an antenna with a magnetic base. This allows quick installation and removal from a motor vehicle or other metal surface.

mA/h –
Milliampere per hour.

MARS –
Military Affiliate Radio System. Established in 1948. These are military affiliated amateurs who provide free communications for overseas GIs and other Federal services.

Matchbox –

"Antenna Tuner". This is an Impedance matching device that matches the antenna system input impedance to the transmitter, receiver, or transceiver output impedance.

MCW –
This is a Modulated Continuous Wave. A fixed audio tone modulates a carrier. This was an older method of sending Morse code.

Megacycles –
"MC". A million cycles per second. This terminology has been replaced by MegaHertz (MHz).

Megahertz –
One million Hertz.

MF –
Medium Frequency - 300-3,000 kHz.

Mic –
"Mike". Short for microphone. The device that converts sound waves into electrical energy.

Microwave –
This is the region of the radio spectrum that refers to frequencies above 1 giga hertz (GHz).

Mixer –

This is a circuit that takes two or more input signals, and produces an output that includes the sum and difference of those signal frequencies.

mW –

Milliwatt (1/1,000 watt).

Mobile –

This is an amateur radio station that is installed in a vehicle.   - A mobile station can be used while in MOTION.

Modem –

This is short for modulator/demodulator. A modem modulates a radio signal to transmit data and demodulates a received signal to recover transmitted data.

Modulate –

To create a radio emission so that it contains information. Either voice, Morse code, music, binary, ascii ect.

Modulation Index –

This is the ratio between the maximum carrier frequency deviation and the audio modulating frequency in an FM transmitter.

MOSFET –

Metal Oxide Semiconductor Field Effect Transistor.

Motorboating –

This is an undesirable low frequency feedback sound. The sound on the audio resembles a motor boat sound.

MUF –

Maximum Usable Frequency. This is a measure of the highest frequency that will support transmissions off of the ionosphere.

Multimode Transceiver –

This is a term for a transceiver that is capable of AM, FM, SSB and CW operation.

mV –

Millivolt (1/1,000 volt).

MW –

Medium Wave. Frequencies in the 300 to 3000 kHz range. It is also used for the AM broadcast band of 530 to 1710 kHz.

N - "NOVEMBER"

NB –

Narrow band. This also can be used for "Noise blanker".

NBFM -

Narrow Band FM.

NCS –

Net Control Station. The station that is responsible for controlling a "Net".

Near Field Of An Antenna –

This is the region of the electromagnetic field immediately surrounding an antenna. This is where the reactive field dominates and where the field strength as a function of angle depends upon the distance from the antenna. It is a region in which the electric and magnetic fields do not have a substantial plane wave character, but vary considerably from point-to-point.

Negative –
A term meaning, “no” or “incorrect”.

Negative Copy –
This is an unsuccessful transmission.

Negative Feedback –
This is a process in which a portion of the amplifier output is returned to the input. It becomes 180 degrees out of phase with the input signal.

Negative Offset –
A repeater’s input frequency is lower than the output frequency.

Net –
A group of stations that meet on a specified frequency at a certain time. The net is organized and directed by a net control station, who calls the net to order, recognizes stations entering and leaving the net, and authorizes stations to transmit.

NiCad –
Nickel Cadmium. This is a type of rechargeable battery.

Nickels –
A term used on DX nets as a signal report 5x5.

NiMH –

Nickel Metal Hydride. This is a newer type of rechargeable battery.

NODE –

A remotely controlled TNC/digipeater. It is used as a connect point in packet radio.

NPN –

A type of transistor that has a layer of P-type semiconductor material sandwiched between layers of N-type semiconductor material.

NTS –

National Traffic System. This is an amateur radio relay system for passing messages.

NVIS –

Near Vertical Incidence Skywave. A type of propagation mode where signals are reflected back down from directly overhead. It is very useful for relatively short distances. By "raining" the signals down from overhead, mountainous regions and the limitations of "skip zones" are overcome.

O - "OSCAR"

OC –
Oceania.

Odd Split –
An unconventional frequency separation between input and output frequencies.

Offset –
Repeaters use two different frequencies. One for transmitting and one for receiving. On the 2 meter ham band these frequencies are 600 khz apart. As a general rule, if the output transmitting frequency of the repeater is below 147 Mhz then the input listening frequency is 600 khz lower. This is called a "negative offset". If the output is above 147 Mhz then the input frequency is 600 khz above. This is called a positive offset.

Ohm –
This is the fundamental unit of resistance. One Ohm is the resistance offered when a potential of one Volt results in a current of one Ampere.

Open Repeater -

This is a repeater whose access is not limited and open to any Ham operator.

OSCAR –
Orbiting Satellite Carrying Amateur Radio.

Oscillate –
To vibrate, generate an AC or other periodic signal.

Oscilloscope –
This is an electronic test device used to observe wave forms and voltages on a cathode-ray tube. It displays time on the X-axis and amplitude on the Y-axis. The Z-axis is intensity of the CRT spot.

OT –
Old Timer. This is someone who has been around Ham Radio a long time.

OTS –
Official Traffic Station

Output Frequency –
This is the frequency of the repeater's transmitter and of the transceiver's receiver.

Over –
A term used during a two way communication to alert the other station that you are returning the communication back to them. Other terms may

be used such as, “back to you”, “microphone to you”. In CW the letter K is used as the invitation to transmit.

P - "PAPA"

P-P –
Peak-to-peak; as in peak-to-peak voltage.

PA –
Power amplifier

Packet Cluster –
A Network of automated packet radio stations for disseminating DX and contest reports.

Packet Radio –
This is a system of digital communication whereby information is transmitted in short bursts. The bursts are called "packets". These packets also can contain call sign, addressing and error detection information.

Paddle –
This is a Morse code key.

Parallel Circuit –

This is a closed circuit in which the current divides into two or more paths before recombining to complete the circuit.

Parallel Conductor Feed Line –

This is a feed line constructed of two wires held at a constant distance apart. It can be either incased in plastic or constructed with insulating spacers placed at intervals along the line.

Parasitic Beam Antenna –

This is another name for the Beam Antenna.

Parasitic Element – This is part of a directional antenna that derives its energy from mutual coupling with the driven element. Parasitic elements are not connected directly to the feed line.

Parasitic –

Oscillations in a transmitter on frequencies other than the desired one; these can produce spurious signals from the transmitter.

PBBS –

Packet Bulletin Board System.

PC –

Printed Circuit.

PEP –

Peak Envelope Power. This is the average power of a signal at its largest amplitude peak..

Peak Inverse Voltage -
"PIV". - the maximum voltage a diode can withstand when it is reverse biased (not conducting).

Phase –
This is the time interval occurring between one event and another in a regularly recurring cycle.

Phase Modulation –
This is varying the phase of a radio frequency carrier in response to the instantaneous changes in an audio signal.

Phone –
Modulating using voice.

Phone Patch –

Sometimes called "Auto Patch". A connection between a two-way radio unit and the public telephone system.

PIC –

PIC is a microcontroller and is short for "Programmable Interface Controller".

Picket Fencing –

This is a condition experienced on VHF and above where a signal rapidly fluctuates in amplitude. This fluctuation causes a sound that resembles rubbing a stick on a picket fence. This can happen when a user's signal is not strong enough to maintain a solid connection to a repeater, such as when passing through or under obstacles while mobile.

Pileup –
Multiple stations calling a DX or contest station at the same time.

PIN –
Positive Intrinsic Negative. Transistor or Diode.

Pirate –
Someone who operates illegally operating on the air.

PL –
Private Line. The same as CTCSS. It is low frequency audio tones used to alert or control receiving stations. It is used to prevent a repeater from responding to unwanted signals or interference.

PLL-

Phase-lock loop

PM –
Phase Modulation, similar to Frequency Modulation.

PNP –
This is a type of transistor that has a layer of N-type semiconductor sandwiched between layers of P-type semiconductor material.

Portable –
A portable station is one that is designed to be easily moved from place to place but can only be used while stopped.

Positive Offset –
The repeater input frequency is higher than the output frequency.

Pot –
Potentiometer. This is a continuously variable resistor. Mostly used for adjusting levels, as in volume control.

Product Detector –
This is a receiver circuit consisting of a beat frequency oscillator with additional circuitry for enhanced reception of SSB signals.

PROM –
Programmable Read Only Memory.

Propagation –

This is the means or path, by which a radio signal travels from a transmitting station to a receiving station,

PSK31 –

A digital transmission mode. Phase Shift keying that uses a 31.25 baud rate.

PTO –

Permeability Tuned Oscillator.

PTT –

Push To Talk. This is the switch in a transmitter circuit that activates the microphone and transmission circuitry,

Pull The Plug –

A term meaning to "shut down the station".

Q - "QUEBEC"

Q –

A figure of merit for tuned circuits. For antennas, the Q is inversely proportional to useable bandwidth, with reasonable SWR.

Quad –

A directional antenna consisting of two one-wavelength "squares" of wire placed a quarter-wavelength apart.

Q-Signals –

Q-Codes. A set of three-letter codes which are used by amateurs as abbreviations. Commonly used on both CW and phone.

QRP –

"Flea Power". This is low power operation. Typically 5 watts output or 10 watts input power.

QSL Manager –

A person, usually an Amateur Radio operator, who manages the receiving and sending of QSL cards for a managed station.

QSO –

A two way conversation.

R - "ROMEO"

RADAR –
Radio Detection And Ranging.

RACES –
Radio Amateur Civil Emergency Service.

Radio Check -
This is a query from a station desiring a report on his/her stations audio quality and signal strength.

Rag Chewing –
Conversing and chatting informally on the Ham radio.

RAM –
Random Access Memory.

R/C –
Radio-control.

Rcvr –
An abbreviation for "receiver".

RDF –
Radio Direction Finding.

Reactance –
This is the opposition to current that a capacitor or inductor creates in an AC circuit.

Reflector –
In antenna applications. It is the element located behind the driven element in a directional antenna. In IRLP applications. It refers to a server that allows multiple Nodes, or repeaters to be linked together at the same time, thus expanding the RF coverage area.

Refract –
It means to bend. Electromagnetic energy is refracted, or bent when it passes through different types of material. Very similar to the way light is bent or refracted when it travels from air into water and water into the air.

Repeater –
A repeater is a receiver/transmitter that listens for your transmission and re-transmits it. Repeaters have the advantage of height and power to extend the range of your transmission. A repeater "repeats" your transmission by listening on one frequency and transmitting on another. The offset is the difference in separation of both the input/listening and output/transmitting frequency.

Repeater Directory –

This is an annual ARRL publication that lists repeaters in the US, Canada and other areas.

Resonance –

A condition where Xc = XL, establishing a resonant circuit - used for selectivity (parallel circuit) or a maximum impedance circuit (series).

Reset -

This is when a repeater timer is reset back to zero. It normally occurs when the carrier of the transmitter drops.

RF –

Radio Frequency. These are electromagnetic emissions that take place in the radio spectrum.

RFI –

Radio Frequency Interference.

RG –

A term describing a type of coax. Such as, RG-8 Coax. RG can either mean "Radio Group" or "Radio Guide".

Rig –

Radio Equipment. Either a radio transmitter, receiver, or transceiver.

RIT –
Receiver incremental tuning. Also known as a Clarifier.

RMS –
Root Mean Square.

Roger –
A term meaning, "I understand".

ROM –
Read Only Memory.

Rotator –
Also called a "Rotor". This is a device attached to an antenna mast which rotates it so that the antenna can point in different directions.

Rover –
A station that operates from several grid squares or counties during a contest

RS-232 –
This is a standard of computer set by the Electronics Industries Association (EIA).

RST –

"Readability, Signal, and Tone". This is a three digit report indicating how well an operator's transmissions are being received.

RTTY –

Radio Teletype. This is a form of digital communications.

Rubber Duck –

This is a flexible, short antenna commonly used on hand held transceivers.

Rx –

Receiver, Receive.

S - "SIERRA"

SAREX –
Shuttle Amateur Radio Experiment. Communicating with the astronauts in space.

SASE –
Self-Addressed Stamped Envelope.

SEC –
Section Emergency Coordinator.

Selectivity –
This is the ability of a receiver to reject signals adjacent to tuned signal.

Sensitivity –
A receiver's ability to receive weak signals.

Series Circuit –
This is an electrical circuit in which the current must flow through every part of the circuit. There is only one path for it to flow.

SFI –
Solar Flux Index

Shack –
This is the area designated for operating the Ham station.

SHF –
Super High Frequency 3 - 30 GHz.

Short Path –
A direct signal path in degrees between two stations.

Signal –
This is a radio emission.

Silent Key –
This is a deceased amateur operator.

Simplex –
This is a communications mode in which a radio transmits and receives on the same frequency.

SINAD –
Signal to noise and distortion ratio.

Single Pole, Double Throw (SPDT) switch -
This is a switch that connects one center contact to one of two other contacts.

Single Pole, Single Throw (SPST) switch –
This is a switch that only connects one center contact to another contact.

SINPO –
This is reporting system used by radio hobbyists to indicate how well a station was received: S=Strength, I=Interference, N=Noise, P=Propagation, O=Overall.

SITOR-A –
Simplex teleprinting over radio system, mode A.

SITOR-B –
Simplex teleprinting over radio system, mode B (FEC mode).

Skip Zone –
This is a "dead zone". Basically it is an area that is too far for ground wave propagation and yet too near for sky wave propagation.

Skyhook –
Antenna.

Sky Wave Propagation –
This is the transmitting of radio waves that reflect off the ionosphere.

SM –

Section Manager.

S-Meter –
Signal Strength Meter.

S/N –
Signal to noise ratio.

Spark Gap –
This is an early transmitter design which used electrical sparks to generate radio frequency oscillations.

Spectrum –
Either the entire or some portion of the electromagnetic spectrum.

Speech Processor –
This is a circuit that increases the average level of the modulating signal applied to a transmitter.

Splatter –
This is interference to stations on nearby frequencies. Splatter can occur when a transmitter is over modulated.

Sporadic E –
Random patches of intense ionization that form in the E-layer of the ionosphere and refract higher frequency signals that normally cannot be refracted by the ionosphere.

SPST –
Single Pole Single Throw. An "On" "off" style switch.

Spurs –
Spurious Signals. These are undesired signals and frequencies in the output of a transmitter.

SQL –
Squelch. A circuit that mutes the receiver when no signal is present, thereby eliminating band noise.

Squelch Tail –
This is a brief bit of noise heard between the end of a radio transmission and the reactivation of the receiver's squelch circuit.

SSB –
Single Side Band.

SSBSC –
Single Side Band Suppressed Carrier.

SSN –
Sunspot Number.

SSTV –
Slow Scan Television.

Straight Key –

This is a non-electronic type of Morse code key with one paddle.

Stub –

A transmission line 'stub' is a length of transmission line that is open or shorted at one end. It is effectively a capacitor or inductor, depending on length, and can be used to achieve a match [VSWR = 1:1] if connected at a selected point in the basic transmission line.

Super heterodyne –

This is a radio receiver scheme which beats or heterodynes a second radio frequency to the incoming radio signals. The combined frequencies form an intermediate (IF) third frequency. This aids in selectivity characteristics.

SW –

Short Wave.

SWL –

Short Wave Listening.

SWR –

Standing Wave Ratio. This is a system of measuring how much radio energy sent into an antenna system is being reflected back to the transmitter.

SWR Meter –

This is device is used to determine the Standing Wave Ratio of an antenna system.

Synch Detection –

Synchronous Detection. This is a method of processing an AM signal to improve audio quality. It also reduces interference from adjacent stations.

T - "TANGO"

TCXO –
Temperature Compensated Crystal Oscillator.

Telegraphy –
This is transmitting information in Morse code format.

Telephony –
This is transmitting information in voice format.

Third Party Communications –
Messages are passed from one Ham to another on behalf of a third person.

Third Party Communications Agreement –
This is an official understanding between the United States and another country that allows amateurs in both countries to participate in third-party communications

TI –
Talk-In Frequency.

Ticket –

This is slang terminology for an amateur radio license.

Timer –

Repeaters may incorporate a timer or transmit limiter to control the length of a single transmission from a user. The time limit is usually set by the repeater owner.

Time Out –

This is when an excessively long transmission on a repeater causes the repeater's timer circuit to stop the transmission.

Tone Pad –

Resembling a standard telephone keypad, it is an array of 12 or 16 numbered keys that generate the standard telephone dual tone multi dialing signals. *(DTMF)*

TOR –

Telex Over Radio.

TNC –

Terminal Node Controller. This is a device interfaces a transceiver to a computer.

TP –

Test Point.

Traffic –

Message or messages sent by radio.

Transceiver –

This is a single unit radio that both transmits and receives.

Tropospheric Ducting –

Propagation of signals above 30 MHz via bending and ducting along weather fronts in the lowest layer of the Earth's atmosphere, the troposphere.

TRX –

Transceiver.

TTL

Transistor Transistor Logic.

TV –

Television.

TVI –

Television interference.

Two Tone Test –

This is a method of testing a side band transmitter by feeding two audio tones into the microphone input of the transmitter and observing the output on an oscilloscope.

Tx –

Transmit. Transmitter.

U - "UNIFORM"

UHF –
Ultra High Frequency 300 - 3000 MHz.

Uncle Charlie –
The FCC.

Unun –
"unbalance - unbalance". This is a device which couples an unbalanced antenna of one impedance to an unbalanced feed line of another impedance.

Uplink –
This is a channel used for earth to satellite communications.

URL –
Universal Resource Locator.

USB –

Upper Side Band. As a general rule of thumb, this is the single sideband operating mode used on the 20, 17, 15, 12, and 10 meter HF amateur bands.

UTC -

Coordinated Universal Time. This is time in the 24 hour format, expressed at the at the 0-degree Meridian, which passes through Greenwich, England.

Utility Stations -

These are stations that are not intended to be heard by the public. They include the marine, embassy, military, aircraft and radiotelephone communications.

V - "VICTOR"

V –
Volt. A unit of electromotive force, or EMF.

VA –
Volt Amperes. This is a measure of apparent power and not True power.

VAC –
Volts Alternating Current.

Varactor Diode –
This is a component whose capacitance varies as the reverse bias voltage changes.

VCO –
Voltage Controlled Oscillator.

VDT -
Video Display Terminal.

VE –

Volunteer Examiner. This is a person who is authorized to administer amateur radio license examinations.

VEC –

Volunteer Examiner Coordinator. This is an amateur radio organization empowered by the FCC to recruit, organize, regulate and coordinate Volunteer Examiners.

VFO –

Variable Frequency Oscillator.

Velocity Factor –

This is the speed at which radio waves travel in a particular feed line. It is expressed as a percent of the speed of light.

VHF –

Very High Frequency 30 - 300 MHz

VIS –

Vertical Interval Signaling. This is the digital encoding of the transmission mode in the vertical sync portion of an SSTV image.

VLF –

Very Low Frequency 3 - 30 KHz

VMOS –

Vertical Metal Oxide Semiconductor.

VOM –
Volt Ohm Meter.

VOX –
Voice Operated Transmit.

VSWR –
Voltage Standing Wave Ratio.

VTVM –
Vacuum Tube Voltmeter.

VXO –
Variable Crystal Oscillator.

W - "WHISKEY"

WAC –
Worked All Continents award from the IARU, administered by ARRL.

Wallpaper –
QSL cards, awards or special event certificates.

WARC –
World Administrative Radio Conference.

WARC Bands –
This is an expression to indicate the bands that were allocated in 1979. They are the 17M, 12M and 30M bands.

WAS –
Worked All States award from ARRL for confirmed contact with each of 50 states.

WAZ –
Worked All Zones award from CQ magazine for confirmed contact with each of 40 zones.

WEFAX –
Weather facsimile. Reconstructed satellite images and photographs.

WFWL –
This is DX-ing term used when the validity of a DX station is in doubt. Work First Worry Later".

White Noise –
This is a term used to describe broad band noises that are generated in a receiver's detector and sampled to control the receiver's squelch. Often a noisy sound.

Wilco –
Will comply.

Wireless –
Using radio as opposed to being wired – telegraph.

Work –
To "work" another station is to communicate with another radio station. A valid two way contact.

WPM –
Words Per Minute. It is the typing speed used in CW or Morse Code.

WX – Weather.

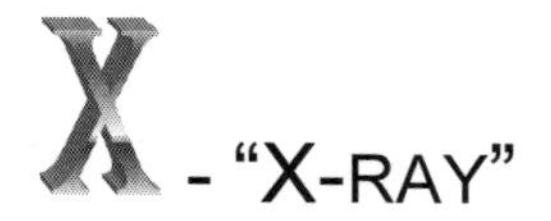

X - "X-RAY"

XCVR-
Transceiver.

XFMR –
Transformer.

XIT –
Transmit Incremental Tuning Control. This allows for slightly changing the transmit frequency while leaving the receive frequency the same. It is useful for split operations.

XTAL –
Crystal.

XVTR –
Transverter. This configures a transceiver to operate on other bands.

XYL –
Ex-Young Lady, wife.

Y - "YANKEE"

Yagi –

In 1926, Hidetsugu Yagi and Shintaro Uda invent the "beam" antenna array. This is a directional antenna consisting of a dipole and two additional elements, a slightly longer reflector and a slightly shorter director. The electromagnetic coupling between the elements focuses maximum power (or reception) in the direction of the director.

YL –

Young Lady. Any female amateur radio operator.

Z - "ZULU"

Zed –

A phonetic for the letter "Z".

Zero Beat –

Adjust the frequencies of two signals so that they are exactly equal and in phase.

Zulu –

Coordinated Universal Time. Also it is the phonetic for the letter "Z".

NUMBERS

.

73 –

Best regards.

88 –

Love and kisses.

807 –

A term used by Hams for Beer. Based on the Beer bottle shape of the 807 tubes of the 1900's.

ABOUT THE AUTHOR

John is a former New York State Correctional Officer turned author. In 2008, John was awarded his General Class “ticket” in Amateur radio giving him the call sign of KC2TAV. John draws upon his life experiences as an officer, husband and father to write books that attract a diverse audience. John lives in Upstate NY, in the foothills of the Adirondack Park, where he enjoys hiking, hunting, gardening and radio communication. He shares his home with his wife, three children and faithful dog, Conan.

OTHER BOOKS BY JOHN PERTELL

The Ham Radio Jargon Quick Reference Guide: Because Understanding Terminology Is Critical Kindle Edition – June 14, 2014 - **ASIN:** B00L04GC1Q

The Q-code Quick Reference Chart For Ham Radio Kindle Edition – September 24, 2013 - **ASIN:** B00FFAYE6Q

The Biblical Greek Alphabet Reference Chart Kindle Edition – June 29, 2013 - **ASIN:** B00DPPAWSW

Made in the USA
Columbia, SC
25 July 2023

20890987R00078